Michael Mosley

TOMORROW'S WORLD

TRANSPORT

B☒XTREE

First published in the UK 1993
by BOXTREE LIMITED, Broadwall House,
21 Broadwall, London SE1 9PL

1 3 5 7 9 10 8 6 4 2

Copyright © (text) Michael Mosley
Copyright Tomorrow's World © BBC 1993 Licensed by BBC Enterprises Limited.
Tomorrow's World logo © the British Broadcasting Corporation 1992 Licensed by BBC Enterprises Limited
Tomorrow's World is a registered trade mark of the British Broadcasting Corporation in the United Kingdom
Trade Mark registration: TOMORROW'S WORLD no B1224487 class 16
All rights reserved

Edited by Miranda Smith
Design by Julian Holland Publishing
Artwork by Tony Gibbons and Steve Seymour
Picture research by Dee Robinson
Cover illustration by Steve Seymour
Cover design by Pinpoint Design Company

1-85283-336-X

Except in the United States of America, this book is sold subject to the condition that it shall not, by way of trade or otherwise, be lent, resold, hired out or otherwise circulated without the publisher's prior consent in any form of binding or cover other than that in which it is published and without a simliar condition including this condition being imposed on a subsequent purchaser

A catalogue record for this book is available from the British Library

Acknowledgements
Boxtree would like to thank Cynthia Page for her help in compiling the Facts and figures section (pp.44-47), and Dana Purvis, the editor of Tomorrow's World for her advice and assistance. They are also grateful to the following sources for help with the artworks and photographs in the book: British Aerospace, Ford, General Motors, Honda, Japan Ship Centre, McDonald Douglas, QA and Rolls Royce.

The photographs that appear in the book were obtained from the following sources:

pp.4-5 The Image Bank; p.6 Shell International; pp.6-7 Chrysler Corporation; p.7 Chrysler Corporation; p.8 Robert Hunt Library; p.9 *above* ZEFA; p.9 *below* General Motors Corporation; p.10 Jaguar; p.11 *above* and *below* Volkswagen; p.14 ZEFA; p.15 (crash sequence) Rover; p.16 Douglas Aircraft Company; p.17 Aviation Week and Space Technology; p.20 British Aerospace; p.23 *above* ZEFA; P.24 *above* East Japan Railway Company; p.24 *below* GEC Alsthom; p.25 Fiat Ferroviaria; p.28 J. Allan Cash; p.29 *above left* ZEFA; p.29 *above right* London Transport; p.29 *below* Urbanizçao de Curitiba S.A.; p.30 Kirk Precision; p.31 *above* David Cannon/Allsport; p.31 *below* Sinclair Research; p.32 *above* J. Allan Cash; p.32 *below* Dartford River Crossing Ltd; p.33 *above* Renault; p.33 *below* Jerrican; p.34 Walker Wingsail Systems plc; p.35 *above* Japan Ship Centre; p.35 *below* JETRO; p.36 *above* Japan Ship Centre; p.36 *below* GAMMA/Frank Spooner Pictures; p.37 Cranfield Institute of Technology; p.40 *left* Steve Brock/Eye Ubiquitous; p.40-41 *above* ETSU; pp.40-41 *below* Tiphook; p.41 Danbrit Shipping Ltd; p.42 Paul Moller; pp.42-43 BEDE Aircraft; p.43 *above* and *below* Toyota; p.44 *left* Anthony Upton; p.44 *above right* Mazda; p.44 *below right* Richard Noble; p.45 *above* Rex Features; p.45 *below* Bob Starr; p.46 Malcolm Heywood; p.47 *above* Debra Lex; p.47 *below* Australian Information Service.

CONTENTS

On the move	4
Making cars less greedy	6
Fuelling the future	8
The intelligent car	10
Tomorrow's car	12
A lethal weapon	14
Taking off	16
The Mega Jumbo	18
Faster than sound	20
Safer than crossing the road	22
Keeping on track	24
The train that flies	26
City transport	28
Pedal power	30
City calming	32
The seas	34
Above the waves	36
The Channel Tunnel	38
Moving goods	40
What is it?	42
Facts and figures	44
Index	48

ON THE MOVE

MAKING PREDICTIONS is dangerous, but one safe prediction about travel in the future is that there will be a lot more of it. Whether by air, sea or on the ground, in tomorrow's world people are going to be travelling more than they do at present. But almost all forms of transport pollute the air, damage the environment and threaten our health. The challenge for the future will be how to move people and freight without making the environment more unhealthy.

This book gives a lot of space to a major environmental villain, the car. This is partly because, in the 21st century, there are going to be many more cars on the roads. An increasing number of them will be electric; they will be 'intelligent', safer and far more fuel efficient. Some cars may even fly. What they will not be doing is going faster – more traffic will mean more traffic jams. Cities will change as they attempt to limit the damage caused by the car. Los Angeles, one of the most polluted cities in the world, is planning to spend $150 billion over the next 30 years on a new public transport system. Switzerland is also developing radical plans to curb car use by making it more pleasant to use public transport than to go it alone.

What is true on the ground is also true in the air: planes, like cars, will face the problems of congestion and pollution. If as is predicted, air travel doubles by the year 2010, we will be spending an increasing amount of time waiting on the ground. Planes will get bigger and safer, some will be faster, and there will be more and more flown by computer.

Unlike other forms of transport, many train journeys will get faster. Already there are several countries with trains that regularly go more than 160 kph (100 mph). Germany and Japan have experimental trains that can levitate and literally fly at speeds of up to 500 kph (300 mph). At least one of these systems is likely to be carrying commercial passengers early in the next century.

This book looks at the many faces of transport at the end of the 20th century and looks forward into the possible world of the 21st . . . .

▷ *An aerial view of highway traffic in California, USA.*

MAKING CARS LESS GREEDY

At the moment in Europe, there are more cars being made every year than babies being born. By early next century there are likely to be twice as many cars and lorries around as there are now. The result will be more pollution unless car-makers can produce 'cleaner' cars. On pages 8-9, we look at alternative fuels, but most cars will be running on petrol well into the next century.

Petrol cars can be 'cleaned up' in many ways. Making petrol lead-free helps, but that only removes one of many pollutants. Catalytic converters to remove much of the remaining pollution from the exhaust will also make a difference. But lean-burn engines, which burn the fuel more thoroughly and have cleaner emissions, may be a better bet in the long term. The best way to manufacture petrol-driven cars that produce less pollution is to make them more fuel efficient.

There are basically three ways to make cars that go further for less fuel: put in a smaller, more efficient engine; make the car lighter; and produce a better aerodynamic shape. The car below left takes those three principles to the extreme, which is why it holds the world record for fuel efficiency. It consumes just 0.0004 litres per km (over 7,500 miles on a gallon of petrol). Not many of us are going to drive around in one of these in the future, but the lessons are there.

△ This record-breaking fuel-efficient car has a speed of 21 kph (15 mph).

Another way to save fuel is to make cars with lighter bodies. Many parts of cars that were formerly made of steel are now plastic, but plastics often lack the stiffness and impact resistance of metals, and they are far more difficult to recycle. Recyclability will be of growing importance in the future, because there will be so many more cars to dispose of. The answer may lie in aluminium. Although it is more expensive than steel, it is about half the weight for the same strength. And, because it does not corrode like steel, it needs no protective treatment, such as galvanizing, which adds to the problems of recycling. Manufacturers have already demonstrated that aluminium cars could easily do 160 km (100 miles) to the gallon.

One idea being pursued by many car companies is a novel type of two-stroke engine. Two-strokes are normally associated with lawn mowers; the only cars to be driven by them in recent times were the East German Trabants, hardly cars of the future. Trabant drivers spent most of their time in a haze of blue smoke because the car ran on a petrol-oil mixture and the burning oil made the exhaust fumes blue. The new two-stroke produces a far cleaner burn by injecting petrol straight into the cylinders under a very precise control. It will give more power for a given size than a conventional four-stroke engine: a 1.2 litre two-stroke is as powerful as a normal 1.6 litre engine today. With a smaller and lighter engine you can have a shorter bonnet, which means a more aerodynamic car. The overall fuel savings should be around 20 per cent. An added bonus for owners is that, because the engine is less complicated, it needs less servicing.

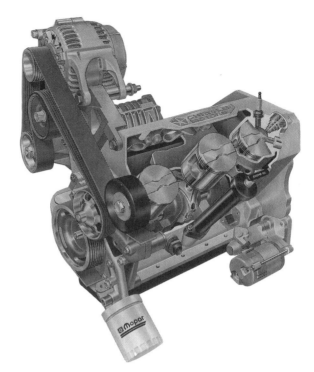

△ The V6 two-stroke engine is only two-thirds of the weight of a comparable four-stroke engine.

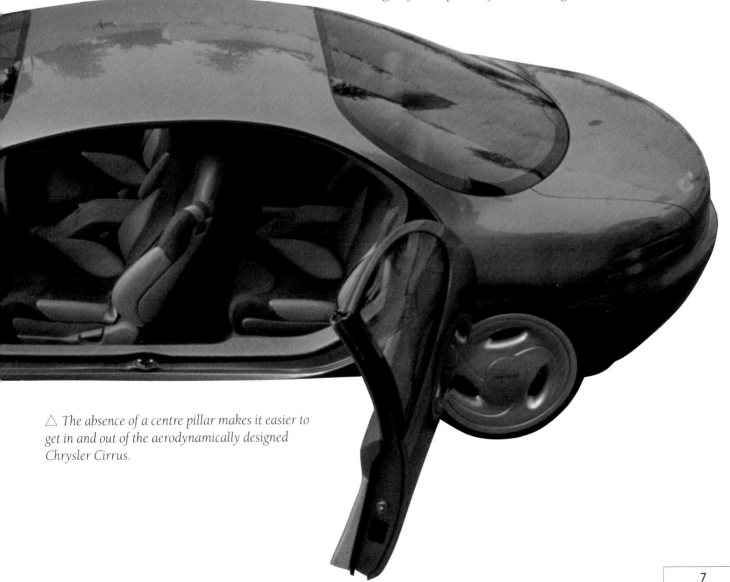

△ The absence of a centre pillar makes it easier to get in and out of the aerodynamically designed Chrysler Cirrus.

FUELLING THE FUTURE

In the 21st century, cars will increasingly be run on fuels other than petrol. This is because petrol-driven cars produce more than 100 million tonnes of atmospheric pollutants every year. These pollutants include:

- **Sulphur dioxide** Responsible for the acid rain that kills trees, and pollutes rivers and lakes.
- **Nitrous oxide** Irritates the lining of the nose; increased levels in towns are believed to be the reason why so many more people today have asthma, hay fever and other allergies. It also destroys the ozone layer.
- **Lead** Causes brain damage and mental retardation.
- **Unburnt hydrocarbons** Can cause cancer.
- **Carbon dioxide** Increased levels of this gas in the upper atmosphere trap heat (the so-called 'greenhouse effect') and threaten to make the world a hotter place.

So what are the cleaner alternatives? The city of Gothenburg in Sweden is planning to run buses powered by natural gas. Alcohol will also become an increasingly important fuel. The Brazilians have several million cars running on alcohol (ethanol) that comes from cane sugar. American, German and Japanese car manufacturers are making vehicles that will run on another form of alcohol – methanol.

One of the alternatives is natural gas (methane). This is not a new idea; there were cars running on methane before the Second World War, and indeed there are now millions of cars and trucks worldwide that run on natural gas.

◁ *This extraordinary vehicle was powered by coal gas carried in the cylinder.*

The French are beginning to introduce a fuel made from rapeseed oil. This makes the car smell a bit like a chip fryer, but the fuel burns far more cleanly than petrol or diesel.

All fuels produce some pollution. So the ideal car is one that does not depend on burning fuel, in other words an electric car. Almost every car manufacturer has plans to sell electric cars, but the GM Impact below is probably the most impressive. Thirty-two batteries give a pair of electric motors enough power to compete with petrol-driven sports models: 0 to 96 kph (0 to 60 mph) in 8 seconds, and a top speed of 160 kph (100 mph). The biggest problem is range. Every 200 km (125 miles), it needs between two and eight hours of recharging.

The Japanese have a car with a similar range, but its batteries can be recharged in 15 minutes. They are made of nickel-cadmium and are very thin, which means they quickly release the heat that builds up during charging. It is certainly an improvement, but what is really needed is a better way of storing electricity.

△ Some British buses will soon be run on fuel made from rapeseed oil.

▽ The GM Impact electric car should be in production by 1994.

An American inventor claims to have developed just that: an improved hydrogen fuel cell. Fuel cells, which produce electric power from hydrogen without burning it, are not new. They were used on the Apollo spacecraft. The challenge has been to make them cheap and convenient enough for everyday use. It is claimed that the new fuel cell literally runs on water. You fill the fuel cell with water and leave it plugged into the mains overnight. The water is converted into a metal form of hydrogen gas (a hydride) by electrolysis. When the car runs, the hydride is converted back into water, producing electricity to power the electric motor. The inventor says his car can run for several hundred kilometres without needing to recharge. Too good to be true? We shall see.

THE INTELLIGENT CAR

To help reduce the increasing traffic congestion which drivers will face in the future, and to improve safety, cars will have to become 'smarter'. There are several electronic guide systems on trial at the moment that can select the shortest route for a journey, avoiding the dreaded traffic jams and road works. With one of them, you tell the system where you want to go, and an arrow immediately appears showing the direction to take.

A computerized map of local streets can be beamed to the car by infra-red beacons, located at junctions around the city. They also send detailed information about traffic jams and hold-ups on the routes ahead. The cars in turn send back information via the beacons to a central computer, which keeps tabs on the current traffic situation.

The trouble with this beacon system is that you need an awful lot of beacons. One alternative is to use a satellite link. The Japanese are the first to fit a system to a production car. It uses radio signals from the Global Positioning System satellite operated by the United States Department of Defense to show drivers where they are. At present it does not give information about traffic jams ahead, or the best way to reach a destination, but within a few years, that should also be possible.

▽ *Roadside beacons transmit information about the best route and traffic conditions to receivers in the cars.*

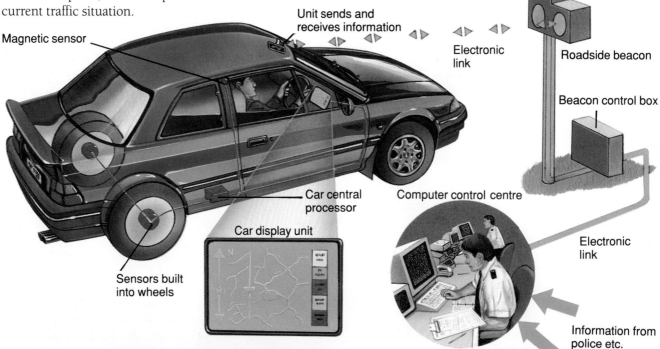

Electronics can do a lot more. This experimental car is equipped with an infra-red camera mounted inside the windscreen to help see through fog. This could be combined with a radar system (also being developed) that detects an obstacle before the driver sees it and shouts an appropriate warning. The whole process might even be made to happen automatically; the radar detects that you are too close to the car in front and puts on the brakes. Or perhaps the radar 'reads' the road ahead and changes the car's suspension to anticipate pot-holes.

◁ *An experimental car using an infra-red camera to see through fog.*

▷ *A 'smart' car like this is beginning to automate the role of the driver.*

The car's central computer could also control the anti-skid brakes, the power supplied to all four wheels and the active steering. Active steering means that the amount the wheels turn when you move the steering wheel depends on the speed of the car; when parking, the wheels turn a lot, and at high speed, very little.

The fully automatic car that can drive itself is not wholly the stuff of science fiction. The Futura above can measure distances, warn if you are too close to other vehicles and even park itself. Other experimental cars can 'see' the side of the road, detect other traffic and use sophisticated computers to steer themselves at speed.

If more cars are to be packed onto the roads in relative safety, the answer might be car convoys. Groups of cars are linked together electronically, and the decision-making is taken almost totally out of the hands of the drivers.

◁ *One driver is in control of the convoy, and could be in any of the cars, although a clear view of oncoming traffic is desirable!*

TOMORROW'S CAR

There are almost as many different ideas for the car of the future as there are different types of car on the road today. But at the start of the 21st century, cars are likely to have some, if not all, of these features. They may be 'hybrids', with both electric and fuel engines. The electric drive would be used in traffic; the fuel engine for longer, faster journeys. The fuel might be petrol or one of the alternatives such as methanol. Most of the features shown already exist in research vehicles, or in the laboratory, and could be on the road by the beginning of the 21st century.

- Satellite receiver
- 'Smart' seatbelts tighten as car slows
- Infrared camera
- Seats automatically adjust to person's shape and sound alarm if driver dozes off
- Smart windscreen wipers that start automatically when it rains
- Central computer controlling active suspension, active steering, non-skid brakes, and power to four-wheel drive
- Automatic fire extinguisher
- Super smooth shape and short bonnet to lessen wind resistance
- Electric engine
- Airbag for front seat passenger
- No-dazzle UV dominant headlamps make road markings fluoresce
- Weather/road sensor
- Lean burn two-stroke engine
- Radar obstacle detector and range finder
- Point to recharge
- Decoder for deciding which engine to use
- Automatic 'active' suspension
- Tyre pressure sensors inside wheels

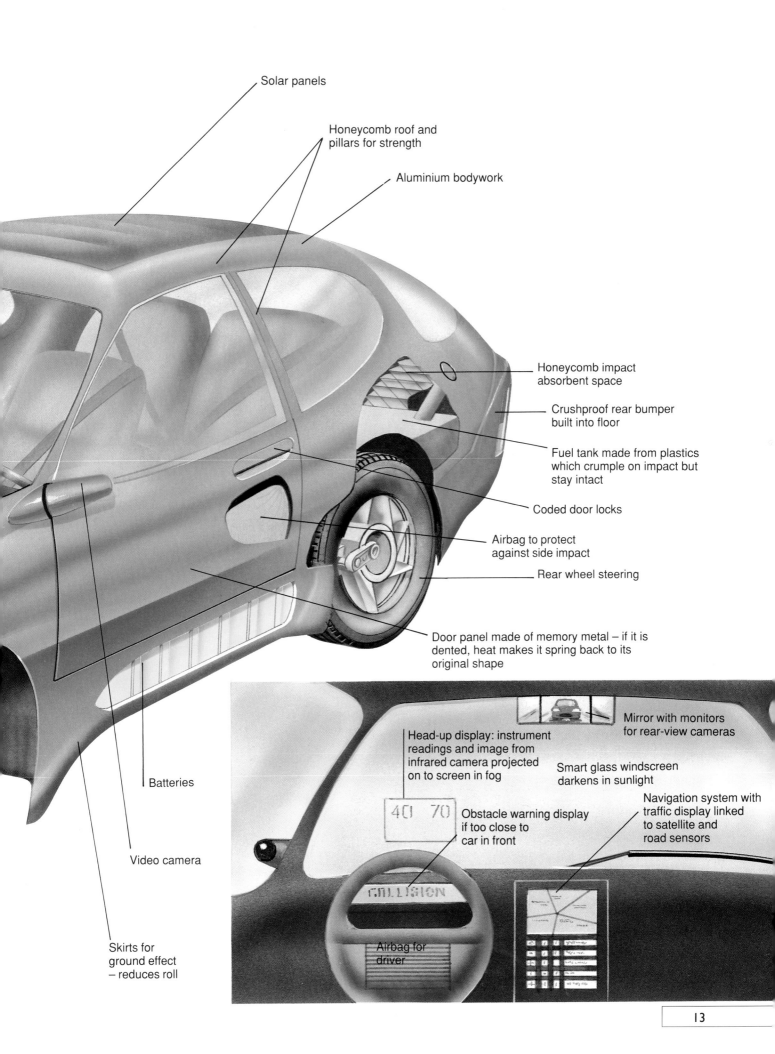

A LETHAL WEAPON

THE CAR is a major killer. It is estimated that 16 million people have been killed worldwide in car crashes – that is more people than were killed by the Black Death. A further 150,000 people die on the roads of North America and Europe every year. Many of the people killed are pedestrians, particularly the very old and young. Improvements in external design to make cars less lethal have been offset by higher vehicle speeds, and it is difficult to see how the number of deaths can be substantially reduced until something is done about curbing the use of cars in towns and cities (see pages 32-33).

In time, car drivers will find more and more of the decision-making is taken out of their hands. Just as in 'fly-by-wire' aircraft, on-board computers will have an override on critical systems such as power, braking and steering. Combined with obstacle detection systems, this should greatly reduce the number of accidents.

Meanwhile motorists and motorbike riders would face less risk of injury if airbags were to be more widely installed. Although they are common in North America, they are not often used in Europe. If the vehicle suddenly decelerates, as in a crash, this is electronically sensed and the airbag is released automatically. The driver is propelled into a pillow of air instead of through the windscreen or over the handlebars. The airbag is inflated by a solid fuel gas generator (a spin-off from space rocket technology) in less than one-tenth of a second. In cars the front passenger can also be protected, and there are even plans to put airbags in door pockets to protect against a sideways hit.

Many deaths and injuries could also be prevented by better use of impact-absorbing materials (for instance in doors), improved seat belt systems and head restraints, and sensibly positioned non-split fuel tanks. To assess how well safety features will perform on the road, it is necessary to know very accurately what happens during a crash. Examining the wreckage of actual crashes, and reports of any injuries received, gives car designers valuable information, but what they really want to know is how much force was necessary to cause the damage.

▽ *Motorway accidents like this are becoming all too frequent.*

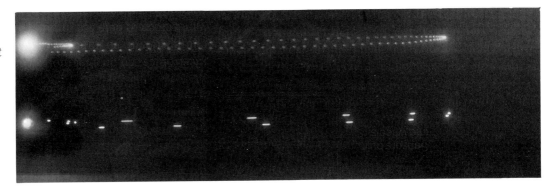

▷ *The film record from a black box for cars, showing the result of an impact.*

To get this sort of information, 10,000 Swedish cars were fitted in 1991 with a so-called 'black box'. It is very cheap and, in a crash, records just one thing – the exact force on the car at the moment of impact. The box holds a photographic film, and a spring attached to a weight, with two lights on top that flash one thousand items a second. In any impact, the weight pulls the flashing lights across the film and they expose a series of dots onto the emulsion. When the film is developed it looks like the picture above; the wider the gap between the dots, the more force was involved in the crash. The project, paid for by car manufacturers, should help designers plan how much impact absorption material is needed, and balance the conflicting requirements of lightness and strength.

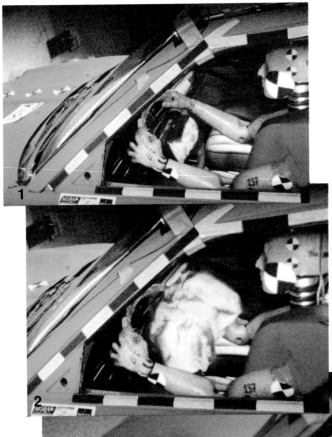

◁ **1.** *In this car, the airbag is positioned in the centre of the steering wheel.*

2. *On impact, the airbag begins to inflate.*

3. *When fully inflated, the airbag cushions the driver from the force of the crash.*

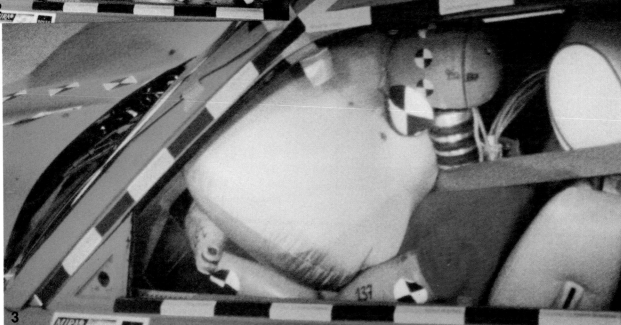

TAKING OFF

THE FUTURE facing air travellers is similar to the prospect for those on the ground: congestion and pollution. The number of people travelling is expected to double by the year 2010. With the number of aircraft movements limited by over-crowding in the air and on the ground, airlines and aircraft manufacturers are thinking big. Huge planes are on the drawing-board (see pages 18-19), capable of carrying up to 1,000 passengers at a time. That compares with today's biggest passenger plane the 747 (the 'Jumbo'), which can carry about 420 passengers.

As there is more flying, so the upper atmosphere will become more polluted. To minimize the damage, aircraft will have to become more fuel efficient. This means more efficient engines, the use of lighter materials to build the planes, and designing them to be more aerodynamic.

Reduced wing drag is not going to come about as a result of changing the shape of planes, because they are already as good as they will get. Instead they can be made more aerodynamic by drilling holes in their wings. This is done to break up the surface of the wing, because a flat surface is not the best shape for reducing drag. Sharks discovered this trick thousands of years ago. They have tiny ridges on their skins to encourage laminar flow over their skin. The same principle has been applied to planes; there are currently several planes flying that have wings pitted with hundreds of millions of laser-drilled perforations. The designers are hoping that this technology could cut fuel consumption by 25 per cent.

▽ *The wing of this test plane is pitted with millions of perforations – the paperclip shows the scale.*

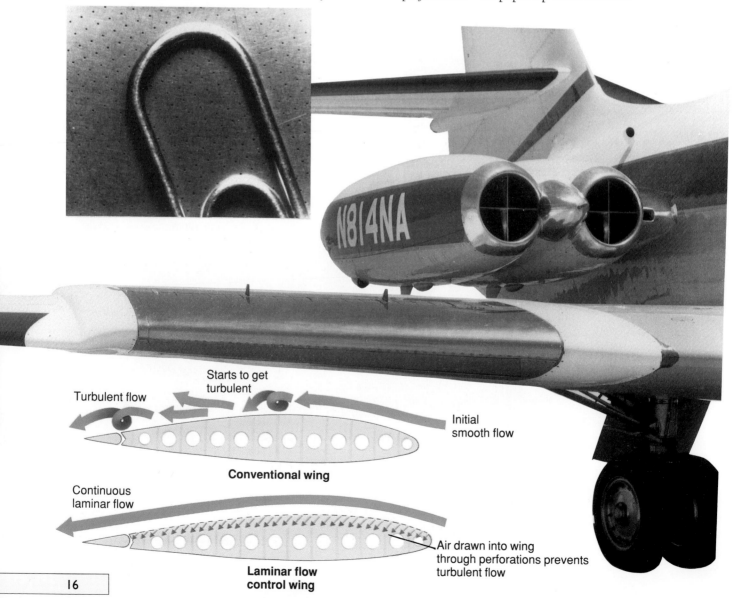

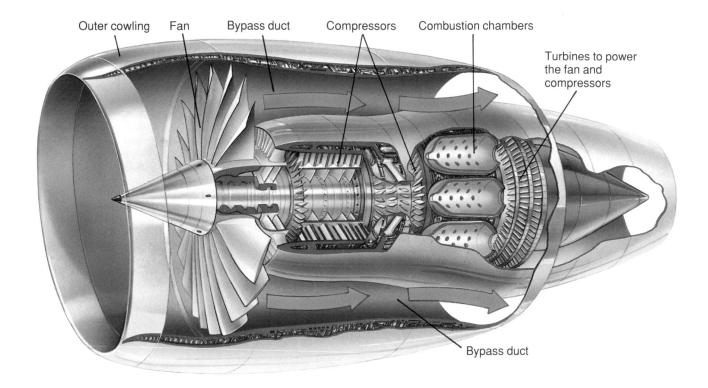

△ *The RB211-535 turbofan.*

Increasing the efficiency of a jet engine involves increasing the amount of what is known as 'bypass'. Not all the air drawn in by the fan at the front of the engine goes into the combustion chamber. Some of the air is propelled along bypass ducting around the engine core before mixing with the hot exhaust gases to provide the thrust.

To get a bigger, quieter, more efficient push you need a bigger flow of air going through the bypass duct. In turbofan engines, which are used now on almost all large passenger aircraft, a large fan at the front accelerates up to five times as much air through the bypass duct, as goes through the engine core. Seventy-five per cent of the thrust is generated by the bypass air. The next generation of turbofans will have fans 3 m (10 ft) in diameter. To make the fans any bigger you would probably have to get rid of the outer cowling, leaving just the engine with huge propeller-like fans in front to accelerate enormous mounts of air. Experimental engines, called propfans, have been built to investigate this idea. As expected, they are very efficient – but also very noisy.

▷ *The propfan of the Yak-42LL has two contra-rotating fans.*

THE MEGA JUMBO

Passenger seat with stereo speakers in the head rest; an in-built video screen provides entertainment and information channels

THE NEXT generation of super-jumbo jets is likely to have two or even three decks. Passengers will be able to exercise, have meals in a central restaurant area, and use computers and telephones linked by satellite to the ground. Business class passengers could have beds to sleep in on the lower deck. However, what the designers are not sure about at the moment is whether the planes would fit into most of today's airports. If the planes are too big then they will have to design wings which can telescope or fold up when the aircraft is on the ground.

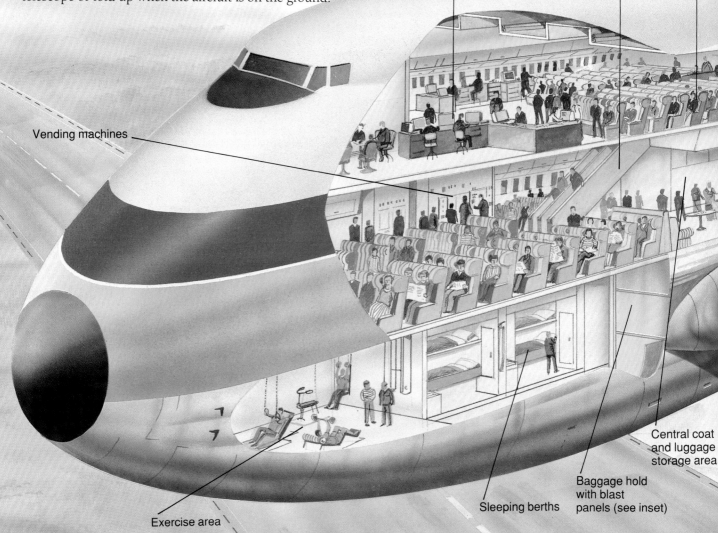

Business centre with computers, telephones and faxes linked to ground and other aircraft by satellite

Seating for about 800 people

Escalators between decks

Vending machines

Central coat and luggage storage area

Baggage hold with blast panels (see inset)

Sleeping berths

Exercise area

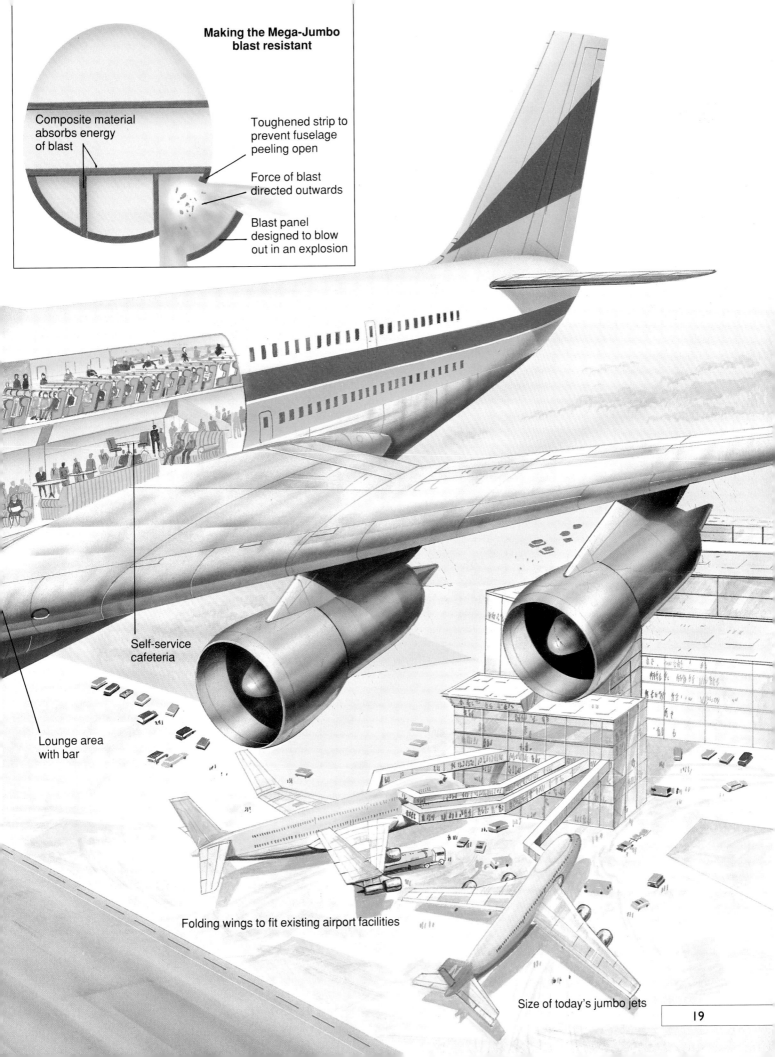

FASTER THAN SOUND

△ *The cost of building a fleet of Super-Concordes may be prohibitive.*

As with subsonic aircraft, the future of supersonic transport is predicted to be in big planes. The current talk is of building a fleet of Super-Concordes, planes with a range of 11,000 km (7,000 miles), able to carry up to 300 passengers at Mach 2. The development costs will be enormous, which is why manufacturers from seven different countries have got together to start the project.

The project will not go ahead, however, unless they can design a plane that is more environmentally friendly than Concorde. One problem is the sonic boom, which could be reduced, but not eliminated, by going for a blunt nose and a very long, thin body. A bigger problem is the engine exhausts, particularly nitrous oxide gases, which are very destructive to the ozone layer. Planes that produce large amounts of pollutants are unlikely to be tolerated in the future.

Supersonic planes have a special problem: to avoid creating a sonic boom over land they have to fly for long periods at subsonic speeds. Concorde can only start flying supersonically when it is 160 km (100 miles) out to sea. Unfortunately, there is no single type of engine that is efficient at both subsonic and supersonic speeds.

As described on pages 16-17, in most modern jet engines only a small percentage of the air taken into the engine actually mixes with the fuel in the combustion chamber and is burnt. Most of the air bypasses the core of the engine; in general, the more that bypasses the core, the more efficient the engine.

A supersonic engine has to be as small as possible, which in practice means pushing most of the air through the engine core. Concorde's engines, for example, have no air bypass system at all, which makes them very inefficient in subsonic flight. It also makes them very noisy.

One possible solution for the future is a variable bypass ducting system. At supersonic speeds, most of the air would pass through the core. But at subsonic speeds extra exhaust nozzles and air intakes would allow much of the air to bypass the core, making the engine much quieter and more efficient.

The dream of the distant future is of a plane, or rather a semi-spacecraft, which would travel halfway round the world in about an hour. It would need even more complicated engines than a supersonic transport. At speeds above Mach 3, the compressors at the front of the engine are no longer needed. The engines would have to convert into ramjets, in which the air is 'rammed' directly into the core by the engine's shape. Hypersonic transports using such hybrid engines to reach speeds of Mach 6 are already on the drawing board.

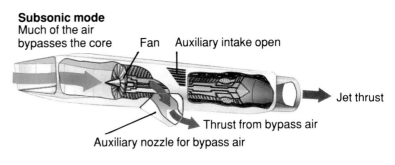

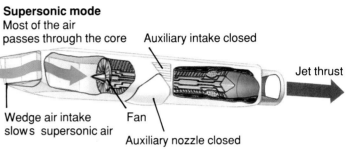

△ *A supersonic engine concept with variable bypass.*

Theoretically, a jet engine may be able to push a plane to near Mach 25, enough to boost it up to 95 km (60 miles) – effectively into space. Then it could continue in a sub-orbital trajectory around the world to land on the other side. Interestingly, because hydrogen is the only fuel capable of powering such an engine, it might be less environmentally damaging than a supersonic plane; when hydrogen is burnt it produces just water. And the cost? Somewhere between enormous and astronomical!

▽ *The hypersonic plane would be capable of travelling from Europe to Australia in 60 minutes.*

SAFER THAN CROSSING THE ROAD

Planes are getting safer. In the 1960s, when you got on an airplane, your chances of being killed were about 1 in 100,000. Now it is about 1 in 500,000. But because air traffic is due to double by the year 2010, the number of fatal plane crashes is likely to increase. There are predictions that by the same year, there will be a major air crash once every other week.

According to a recent survey, half of all fatal crashes could be prevented if full use were made of two standard items of safety equipment. The first is called a Ground Proximity Warning System. It sounds an alarm if the plane is getting too close to the ground. Most planes are fitted with it, but in several recent fatal crashes, the pilots failed to respond to the alarm – the equipment has, until quite recently, been unreliable, so its warnings are sometimes ignored. Over the last few years the systems have been made much smarter: they can take account of the plane's speed, and intended rate of descent, before deciding if there is any danger. You can even tell the system what the terrain is like around the airport you are heading for, and it will adjust its sensitivity accordingly. If all airlines bought the latest equipment and properly trained their crews in its use, it is claimed that about 500 deaths a year could be avoided.

The other equipment that could drastically cut crashes is the instrument landing system (ILS). It makes landing safer by helping pilots align their aircraft with the centre of the runway. Radio pulses are sent from the ground to the plane, effectively providing a beam which the pilot can fly down. The information can also be fed directly to the automatic pilot which can land the plane itself. By 1998, ILS is due to be replaced by an improved system using microwaves called MLS. MLS systems can guide the plane in along a curved flight path, and are less affected by surrounding high ground. However, more than half of the airports in the world have not yet installed either ILS or MLS. Many are hoping that airlines will soon install satellite navigation systems. These systems are so accurate that in theory they could guide an aircraft in to a perfect touchdown, with little or no equipment needed on the ground.

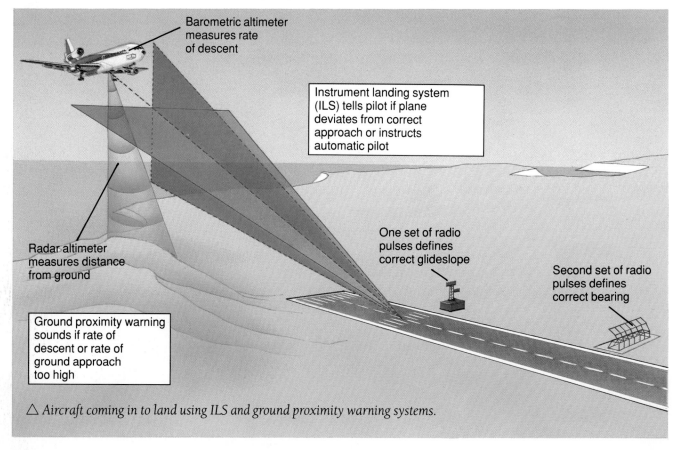

△ Aircraft coming in to land using ILS and ground proximity warning systems.

Although automatic systems do sometimes make mistakes, most experts are convinced that the elimination, as far as possible, of human error will make flying safer. Gradually, electronics are taking over in the cockpit; the pilot of a modern airliner is in effect 'flying' a battery of computers.

In fact, some people have suggested that it would be better to do away with pilots altogether. But that prospect is a long way off. Not only are pilots needed to cope with the unexpected, but most passengers still prefer the idea of having a human in the cockpit!

▽ *One by one, all the instruments and controls in an airliner's cockpit are being taken over by computers.*

In most aircraft accidents fire is the great enemy. When a plane crashes, the spread of fire is usually very rapid, and even if fire tenders arrive quickly, pouring foam on the outside is not the best way of cooling the inside, where the people are. It is even possible that the foam actually makes the inside hotter by creating an insulating layer. One idea which is gradually gaining acceptance is the introduction of on-board water sprinklers. These use very small amounts of water to produce a fine mist, which cools the air inside the cabin and prevents the fire spreading as fast as it would otherwise. When fire tenders arrive, they can connect their equipment to the sprinklers to fight the fire from the inside.

▷ *It may not be long before planes are equipped with internal sprinkler systems.*

KEEPING ON TRACK

△ The central carriages of the new Bullet trains are double decker.

▽ The TGV (Train à Grande Vitesse) on the track.

THE GREATEST challenge to planes, at least over distances of up to 1,000 km (625 miles) comes from the new generation of high speed trains. These trains are capable of speeds of over 250 kph (150 mph), and run on specially upgraded track with fewer curves. The first of these trains was the famous Bullet train in Japan, and a new double-decker version has just been unveiled.

In Europe, several countries have high speed lines in operation. The French TGV has attained an astonishing 515 kph (320 mph) in a test run, and in the future should reach speeds of 400 kph (250 mph) in normal service. In 1989, a plan was put forward by the twelve national rail companies of the European Community to develop a huge, high speed network linking Europe's chief towns and cities by the year 2015.

Building this European network will be very expensive, and much of it may never be completed. Sceptics doubt that high speed rail services could be commercially viable except where they link major cities such as London, Paris, Brussels, Amsterdam and Munich. Most of these journeys would be cut to less than two hours.

Just as on the road and in the air, the risk of accidents to future high speed trains will be reduced by taking critical decisions away from the driver. To make sure the driver does not speed recklessly, the trains will all be fitted with protection systems, linked to computerized signals, which will override the driver who does not obey the signals or keep to the speed limits.

▷ *The high-speed railway network for Europe currently under discussion.*

△ *Il Pendelino travels fast round curves in the track.*

An alternative approach to faster rail travel is to develop trains capable of sustained high speeds on existing curvy track. The Italians have a train called Il Pendelino, or 'the train that tilts', for going fast round curves. British Rail first began work on this concept in the 1970s, but the so-called Advanced Passenger Train was eventually abandoned. Il Pendelino is currently in service. As the train goes into a bend, an on-board gyroscope and accelerometer (which measures centrifugal force) provide information to a computer, which decides how much tilt is needed to compensate. Powerful hydraulic cylinders tilt each carriage just the right amount. The tricky part is to control the tilting movement so that passengers do not get motion sickness, and drinks do not get spilt.

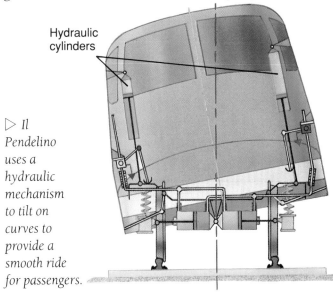

▷ *Il Pendelino uses a hydraulic mechanism to tilt on curves to provide a smooth ride for passengers.*

THE TRAIN THAT FLIES

IN AMERICA, a consortium is planning to invest $600m in a unique new high speed railway track in Florida, running just 14 km (nearly 9 miles) from Orlando airport to Disney World. The track will carry a train that literally flies at speeds of over 400 kph (250 mph), and is scheduled to go into service in 1996.

For 30 years, researchers have been working on these trains which use the principle of magnetic levitation (Maglev). The Japanese and Germans have both built working prototypes. Recent successful high speed runs have convinced the Americans to back the German system.

▽ *The German Transrapid 06 at speed on the track.*

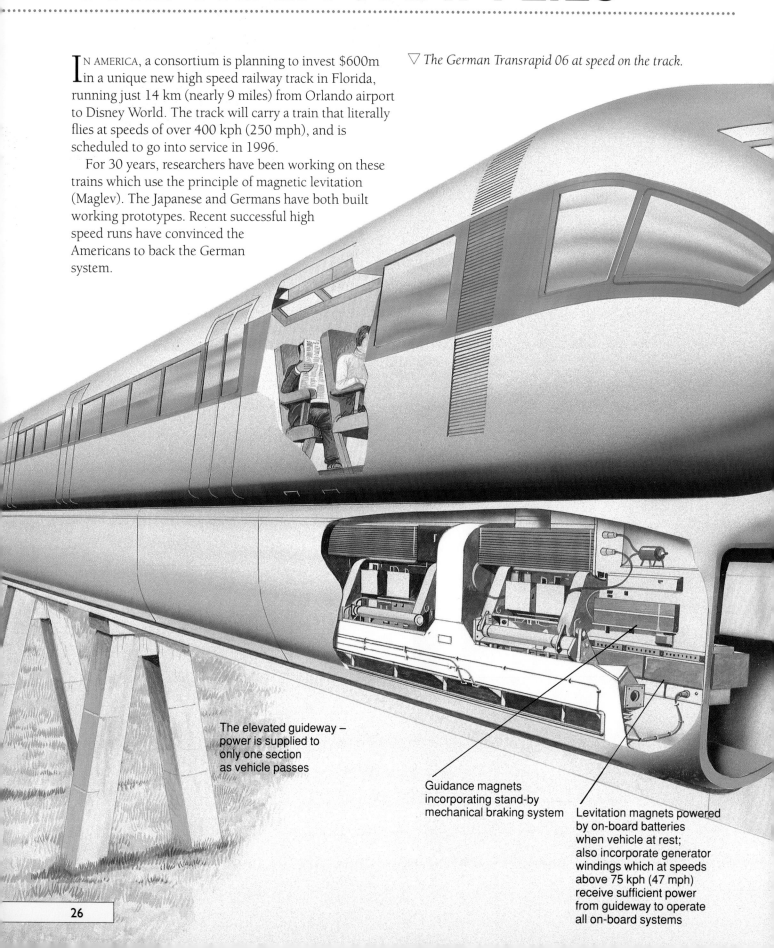

The elevated guideway – power is supplied to only one section as vehicle passes

Guidance magnets incorporating stand-by mechanical braking system

Levitation magnets powered by on-board batteries when vehicle at rest; also incorporate generator windings which at speeds above 75 kph (47 mph) receive sufficient power from guideway to operate all on-board systems

Guidance and stand-by brake rail

Coils in guideway generate travelling field to propel vehicle forward and for braking; coils also transfer power to the vehicle for levitation etc.

Steel rail for levitation

THE GERMAN MAGLEV

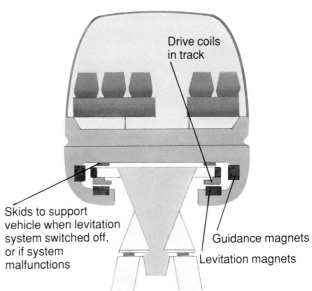

Drive coils in track

Skids to support vehicle when levitation system switched off, or if system malfunctions

Guidance magnets

Levitation magnets

The German train levitates by attraction. The magnets in the overhanging part of the train are attracted up towards the track keeping the train 1 cm (0.4 in) or so above the surface. Magnetic forces are also used to propel the train forward.

THE JAPANESE MAGLEV

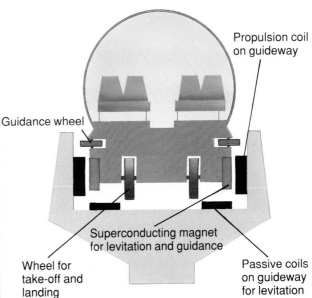

Propulsion coil on guideway

Guidance wheel

Superconducting magnet for levitation and guidance

Wheel for take-off and landing

Passive coils on guideway for levitation

The Japanese train levitates by repulsion. Powerful superconducting magnets in the train induce an opposing field in the track that pushes the train 10 cm (4 in) up into the air. This vehicle currently holds the world speed record for a train of 517 kph (323 mph).

CITY TRANSPORT

For much of the time in London, the average speed of a vehicle travelling around town is less than it was in Victorian times. All over the world, from Los Angeles to Singapore, city dwellers have realized that improving the quality of life will mean reducing congestion. This can be done by curbing the use of the car and improving public transport. Underground subway systems are phenomenally expensive, and in the future few cities will be able to afford them – but there are other options.

The city of Zurich, in Switzerland, has one of the best public transport systems in the world; much of it is built around the tram. Light railway systems (trams) which run at street level are much cheaper to build than an underground train network. If you want to stop the trams getting snarled up in traffic, then you have to do what they do in Zurich, and give them priority. There are induction loops at every intersection. When a tram goes by, it triggers a change in the traffic lights so that the tram does not need to stop. A central computer knows where every tram is in the city, and keeps tabs on which trams are ahead of schedule and which behind, so the drivers can be told to go faster or slower. In Zurich, you need never wait longer than six minutes for a tram, so many people use the service and there is relatively little congestion. Many cities around the world either have or are building similar systems. Some, like Manchester in England, are reinstalling networks that were ripped out in the 1950s.

▽ *A tram on track in the city of Zurich.*

◁ *A People Mover in Sydney, Australia.*

Another idea is the so-called 'People Mover'. This consists of a large number of small electric trains, under computer control, travelling round a city on a guideway which is often raised above ground level; Southampton will be the first city in Britain to have such a system. Trains run on rubber rather than steel wheels, and make much less noise than trams. Because they are computer controlled, in peak periods trains can follow each other at intervals of less than a minute. People Movers are already in use in several cities around the world. A system in Lille, France, is now used daily by about 100,000 people.

The cheapest way of improving public transport is to do something about the buses. Trials have started in London and in Augsberg, Germany, on bus stops with electronic signs which tell you how long you are going to have to wait for the next bus. What they cannot do is tell you if the approaching bus is full.

Britain's old-style double-decker buses look good on postcards, but a service using smaller, streamlined buses that ran more rapidly and frequently would probably be more efficient. The town of Curitiba in Brazil has been built up around a system of dedicated fast bus-ways. Throughout the city buses have priority over cars. The system includes the world's biggest buses, which can carry 270 passengers, and moves four times as many people as Rio de Janeiro's expensive underground system.

△ *An automated bus-stop in the town of Augsberg, Germany.*

△ *Dedicated bus-ways in Curitiba, Brazil.*

PEDAL POWER

THE FASTEST, cheapest and least polluting way of getting around most major towns is the bicycle. The world's 800 million bikes outnumber cars two to one, and transport far more people. Bicycles are already so well suited to what they do that the best way to improve them is probably to find ways of making them out of lighter materials.

Bicycles with frames made of magnesium are now appearing in the shops. These are lighter and stronger than conventional frames – and not a great deal more expensive. In fact, magnesium is one of the world's most abundant metals; each frame uses the magnesium extracted from $1^1/_2$ cubic metres of sea water.

Most improvements to the bike have focussed on the sub-systems like brakes and gears. One idea, now incorporated into top of the range bikes, is to change the shape of the chain wheel from round to oval. In fact, this is not a new idea – it was first tried some sixty years ago – but, now that many bicycles are built with a choice of two or even three chainwheels of different sizes, it may become popular. Cyclists could have at least one of their chainwheels oval. Its advantage is seen when pedalling slowly, for instance uphill. Anyone who has done this will know that the point when you have to push hardest is the 'dead-spot' when one pedal is directly above the other. An oval chainwheel effectively gives a lower gear at this point so that the pedals move faster through the dead-spot, and gives a higher gear when you can push productively.

△ *Phil Anderson, a member of the Dutch TVM team, riding the Kirk magnesium-framed bike in the 1990 Kellogg's Tour of Britain race.*

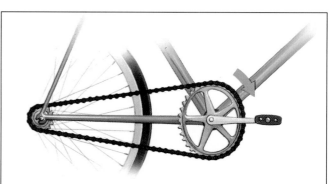

Power stroke: the oval ring acts like a large chainwheel giving a higher gear; the pedals move more slowly.

Dead-spot: the oval ring acts like a small chainwheel giving a lower gear; the pedals move faster.

This bike helped Chris Boardman win an Olympic gold medal in the 1992 4,000 metres individual pursuit. At £4,000 a time it is unlikely that many of us would want to pedal around on one, but the manufacturers hope to produce a cheaper production version! The frame, made of carbon fibre, is supremely aerodynamic, with almost every leading edge shaped like an aerofoil even the handlebars are 'winged'. Not all the ideas in it are new; the front fork has been replaced with the help of an axle that the designer copied from a bicycle originally built in 1897.

▷ *Chris Boardman rides the extraordinary Lotus bike to victory in the 1992 Olympics in Barcelona.*

A development which could take some of the hard work out of cycling and make bicycles even more popular is the electric bike. Like the electric car, there have been electric bikes around for some time; and like the car, the problem has always been the batteries. With a bike, weight is clearly crucial, so the key to success is to keep down the weight of both the batteries and the motor. The latest version, developed by the inventor Clive Sinclair, uses nickel cadmium batteries. Unlike the normal car battery (lead-acid), they are relatively light and can be recharged fairly fast (within about an hour). The batteries are built into the frame of the bike, which itself is made of very light aluminium alloys and glass fibre composites. They power an electric motor, which is exceptionally light because its magnets are made of an alloy of iron, boron and neodymium. The result is a bike that weighs about 11 kg (25 lbs) – roughly the same as an average racing bike – and which will run at 19 kph (12 mph) for about an hour, or much longer if you do your bit with the pedals.

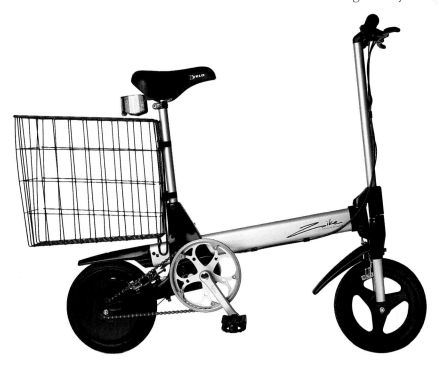

◁ *Clive Sinclair's electric bike, the Zike.*

CITY CALMING

BETTER PUBLIC transport and encouraging bicycle use are not in themselves going to make cities nicer places in which to live. If we want to cut deaths on the road and reduce pollution, we may have to take more active measures to minimize traffic in our towns. In different parts of the world different schemes are being tried.

In Japan, drivers in Tokyo have to prove that they have access to off-street parking before they are allowed to buy a car. In France, the authorities in Paris have tried to cut the number of legal parking spaces by 200,000 to under half a million, but still cars pour into the city centre. In Los Angeles, companies are given a financial incentive to persuade their staff to set up car-share or mini-bus services. On many freeways, lanes are reserved for buses and shared cars. In Singapore, drivers have to buy a ticket before they are allowed to drive into the centre of the city.

▷ *A cycle-designated lane in the town of Alkmaar, Holland.*

And in Oslo, Norway, they have gone one step further. Drivers must buy a card, like a phone card, and as they drive into the centre of the city, roadside microwave beacons automatically subtract a unit from the card. If the driver has no card, the beacon takes a picture of the car's number plate and the driver is automatically fined. The money raised by this method has been used to build an enormous tunnel under Oslo. Traffic wanting to go from one side of the city to the other can go under, rather than through. In England, drivers using the Dartford Tunnel under the river Thames are charged a toll electronically by a very similar system. And in a trial in the London borough of Richmond, electronic meters have been fitted to 100 cars and linked to a network of transmitters spread through the area. The idea is that it will operate as a 'pay as you go' system – the more you travel, the more you pay.

◁ *A microwave beacon extracts a toll from travellers at the Dartford Tunnel.*

A novel type of park and ride using an amazing shrinking electric car is planned in France by 1995. The idea is to drive to the outskirts of a large town or city, leave your petrol-driven car there and rent this tiny electric car. Parking in the smallest spaces is then made easier by the fact that this 2.6 m (8 ft 6 in) car can contract to just over 2 m (7 ft).

▷ *The Zoom will travel 152 km (95 miles) in town traffic on one single charge at speeds of up to 120 kph (75 mph).*

◁ *The body length of the Zoom, also nicknamed 'The Grasshopper', is shortened by pulling in the back wheels.*

The easiest and most effective way of reducing traffic in cities is to pedestrianize city centres. That means putting up physical barriers so that cars cannot be driven there. Studies have shown that when this is done, businesses flourish, more people want to move back into the centre of town, and damage to buildings from pollution drops dramatically.

Dramatic effects can be achieved on a more limited scale, by narrowing roads and adding humps to reduce average speeds. It has been estimated that the enforcement of a 32 kph (20 mph) speed limit in British cities would save a third of the 4,700 annual road deaths and the 47,450 serious injuries.

◁ *A pedestrianized precinct in Bordeaux, France.*

THE SEAS

SHIPS ARE unlikely ever to regain their importance as a way of carrying people around the world, but for a long time they will remain the most important method of transporting freight. And like others in the transport business, shippers are looking at ways to improve their fuel efficiency and reduce costs.

One way of achieving lower fuel consumption is to go back to sailing ships. The sails in the picture below, however, do not look like conventional ones. They are computer controlled to derive the maximum benefit from any wind, and are very simple to operate. Wingsails have been on trial on this tanker for several years, successfully surviving three hurricanes with winds of over 100 knots, and substantially reducing the fuel costs.

▽ *Tankers and yachts have been equipped with Wingsails.*

In Japan, they are convinced that the economic future of long haul shipping will depend on unmanned, computer controlled 'intelligent' ships. A 50 m (165 ft) training ship, the Shioji Maru, has already completed successful trials with the on-board computers steering it along a predetermined course, using radar to avoid other shipping, and responding to new instructions relayed by satellite. The ship successfully navigated its way through a congested harbour, and demonstrated fully automatic docking. Eventually, flotillas of unmanned vessels could cross oceans together, accompanied by a manned mother ship in case any mid-ocean maintenance is needed.

▷ *In Japan, designs already exist for new cargo ships much like the bulk carrier Pacific Pioneer, but controlled by computer. These ships could be at sea within five years.*

Below has to be the most futuristic ship on the seas at the moment. It has no rudder or propellor, but instead is magnetically steered and powered. The idea was first thought of in the 1960s, but the Japanese are the first known to have built a prototype. There are rumours that the Russians may also have at least one magnetically powered submarine silently cruising the depths.

Magnetic drive is based on the principle that if you pass a current through a fluid in the presence of a magnetic field, the fluid will be forced away from the magnet. In the Japanese ship, Yamato-1, the sea water around the ship is first ionized by a powerful current to produce charged particles, particularly chlorine ions. Then a superconducting coil in the hull produces a very powerful magnetic field. This produces a force on the chlorine ions that drives the ship forward. Altering the field steers it left and right. The ship has demonstrated that this method is a more efficient way of transforming power into movement than using a propeller. The designers suggest that in the future magnetic ships should be capable of 50 knots and high temperature superconductor technology could allow them to be built at a reasonable cost.

There are potential problems, however, that Yamato-1 may help to solve – such as how to make sure passengers are properly protected against the powerful magnetic fields. Also, the hull must be prevented from attracting bits of metal that the ship passes.

▽ *The prototype Yamato-1 has achieved about 8 knots.*

ABOVE THE WAVES

ONE OF the few growth areas in shipping has been driven by the demand in Japan, Europe and other parts of the world for more comfortable, high-speed ferries. This has led to several innovative ideas for lifting the main body of the ship's hull out of the water and above the waves.

▽ *The Japanese 'Super Shuttle 400'.*

The current trend is towards catamaran ferries – one, the SeaCat (below), holds the record for crossing the Atlantic. They are far more stable and much faster than conventional boats, their twin hulls effectively cutting through the waves rather than riding on top of them. Unfortunately, as with the hovercraft, many people find the vibrating motion still makes them seasick.

Other designers have continued to develop hydrofoils which have been around for a long time. The first one was invented accidentally by an Englishman, Thomas Moy, in 1861. In order to study the aerodynamics of an aircraft, he stuck a pair beneath his boat. He was very surprised to discover that his boat began to 'fly'. Putting hydrofoils on a catamaran could produce a large, high speed, very stable ferry. Several are now on the drawing-board. On this one (left), designed to carry 350 passengers at 40 knots, the two hulls of the catamaran are connected by fully submerged hydrofoils.

▽ *A SeaCat holds the record for the fastest Atlantic crossing (see page 47).*

These boats may soon be joined by an even more unusual craft. It is called the Caspian sea monster. It was given that name by Western intelligence experts who studied satellite photographs of this huge plane – 100 m (328 ft) long, 40 m (131 ft) wide – flying extremely low over the Caspian Sea. Like a hovercraft, the craft can skim across the water riding on a cushion of air. But it is much faster than a hovercraft and it can also climb like a plane. It works by what is known as the 'wing-in-ground effect'. This is experienced by all planes as they come in to land. A few metres above the land or sea there is an added lift to the wings caused by a cushion of air trapped beneath the plane.

Until the summer of 1992, the Russian craft was a closely guarded military secret, and used mainly for ferrying troops. Now, however, Russian scientists are collaborating with Western companies to turn this military plane into a commercial venture. A commercial version would use additional engines, and flaps on the wings, to generate a cushion of air beneath the plane even when it was stationary. In this way, it could hover, land and take off in a very short space.

At speed, vertical vanes beneath the wings would help to maintain the cushion of air. The Russians estimate that, cruising at a height of about 14 m (45 ft) above the water, it would use about one-fifth as much fuel as that used by a normal high-flying plane.

Although designed for longer journeys, one suggestion is that it could be used to ferry up to 300 people or 30 tonnes of cargo across the English Channel, doing the trip in about 10 minutes. Other countries such as South Korea and Indonesia are interested in using the craft as an island hopper. There is one problem that needs to be solved. It is said that if it travels over a rough sea it feels like travelling over cobblestones in an old-fashioned coach. The Russians are currently working on an automatic flight control system which they say should improve the quality of the ride.

△ The Caspian sea monster in action on the Black Sea.

▽ An artist's impression of the proposed wing-in-ground effect craft.

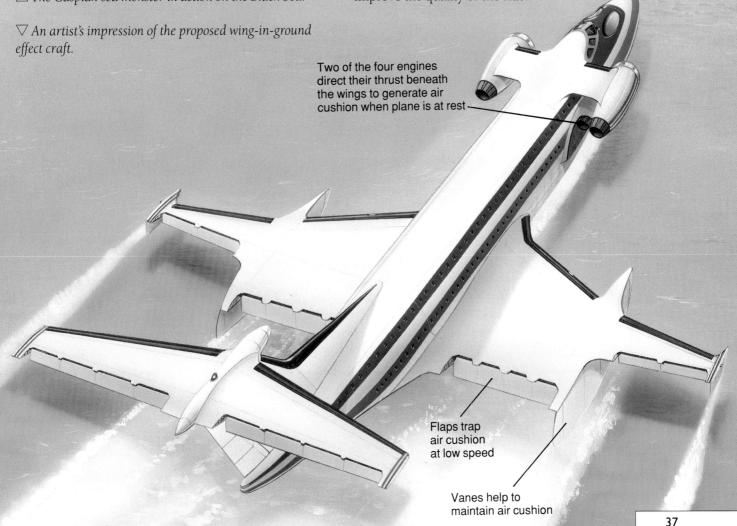

Two of the four engines direct their thrust beneath the wings to generate air cushion when plane is at rest

Flaps trap air cushion at low speed

Vanes help to maintain air cushion

THE CHANNEL TUNNEL

IN FUTURE, a lot of traffic crossing the English Channel will go *beneath* the water. The tunnel was designed to cope with as many as 600 trains a day in each direction running at speeds of up to 160 kph (100 mph). The trains include 'shuttles' carrying road vehicles back and forth between Folkestone in England and Calais in France; intercity passenger trains linking London with Paris and Brussels; night sleepers covering longer distances; and trains moving freight between all parts of Britain and mainland Europe.

The tunnel forms an important link in the proposed European high speed rail network. The French TGV line to the tunnel is already in place, but work on a dedicated high speed line at the British end is unlikely to be completed before the year 2000.

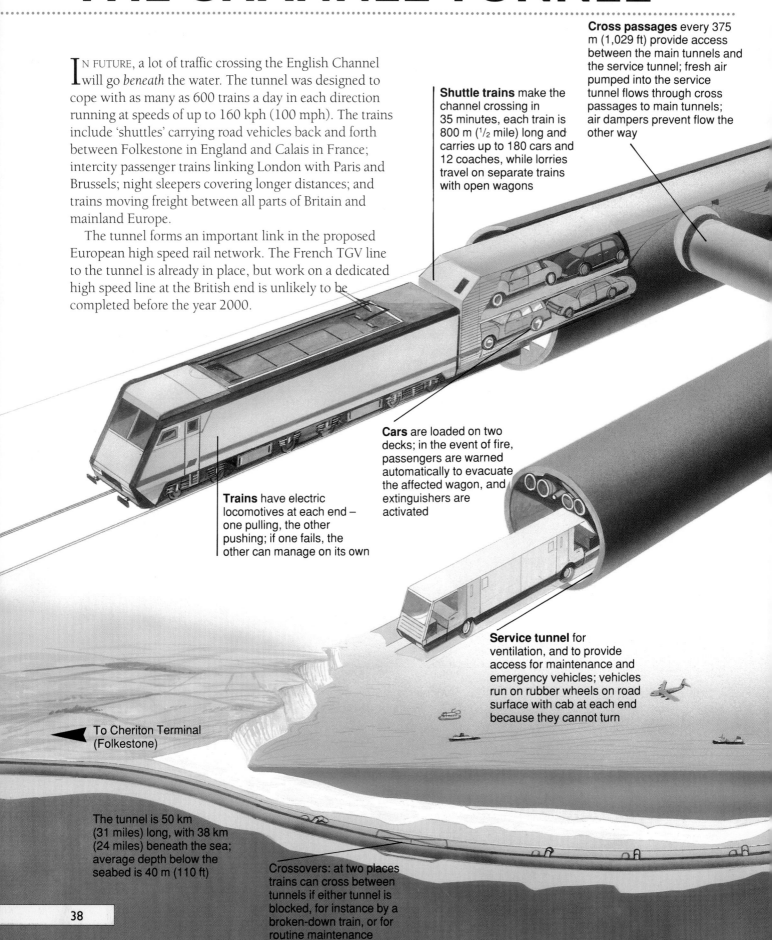

Cross passages every 375 m (1,029 ft) provide access between the main tunnels and the service tunnel; fresh air pumped into the service tunnel flows through cross passages to main tunnels; air dampers prevent flow the other way

Shuttle trains make the channel crossing in 35 minutes, each train is 800 m (1/2 mile) long and carries up to 180 cars and 12 coaches, while lorries travel on separate trains with open wagons

Cars are loaded on two decks; in the event of fire, passengers are warned automatically to evacuate the affected wagon, and extinguishers are activated

Trains have electric locomotives at each end – one pulling, the other pushing; if one fails, the other can manage on its own

Service tunnel for ventilation, and to provide access for maintenance and emergency vehicles; vehicles run on rubber wheels on road surface with cab at each end because they cannot turn

To Cheriton Terminal (Folkestone)

The tunnel is 50 km (31 miles) long, with 38 km (24 miles) beneath the sea; average depth below the seabed is 40 m (110 ft)

Crossovers: at two places trains can cross between tunnels if either tunnel is blocked, for instance by a broken-down train, or for routine maintenance

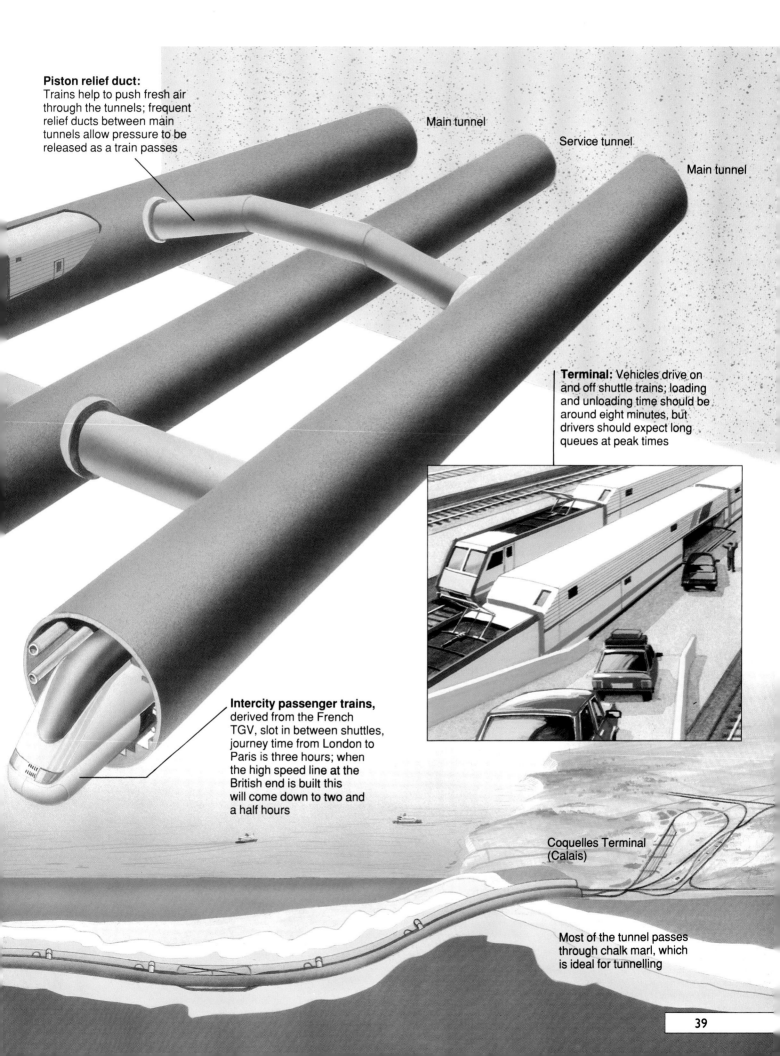

MOVING GOODS

The last few years have seen a dramatic increase in the amount of freight being transported from one place to another. This has been accompanied by a growth in the number of lorries on the roads – a growth that is likely to continue.

Recent trials have shown that streamlining lorries cuts fuel consumption by an impressive 25 per cent and pays for itself in about two years. The trials used 17 tonne trucks that look like shoeboxes on wheels. Fibreglass mouldings are used to produce the streamlined effect. The most important modification is adding a collar between the cab and the body of the truck, which enormously reduces turbulence and drag.

Lorries are potentially a worse health hazard than cars. It is possible that by using fuels other than diesel or petrol, lorries could be made 'cleaner', but they will still clog up the roads and cause other types of environmental damage. For example, they make up about 5 per cent of traffic, but make more noise than all the rest, and a fully loaded lorry does as much damage to the surface of a road as 60,000 cars.

If freight is transported by rail instead of road, there are many advantages as can be seen from the following figures, calculated for each tonne carried 1 km.

	Lorry	Rail
Carbon dioxide emissions	0.22 kg	0.05 kg
Nitrous oxide emissions	3.6 g	0.22 g
Hydrocarbon emissions	0.81 g	0.05 g
Soot	0.27 g	0.03 g
Injured people	248	10

(for each billion tonnes carried 1 km)

△ Lorries are beginning to dominate the roads in the western world.

▽ *To improve fuel efficiency, the collar streamlines this lorry to cut air drag.*

▽ *A tiphook lorry on the road in San Francisco, California.*

Part of the reason very little freight goes by rail is that it is more expensive, but there is also a problem with loading and unloading bulky goods. This is one solution. It is called a tiphook trailer. The body of the truck is winched up onto a rail wagon, then moved by train to a convenient rail head, where it links to another 'front half' and drives away. Loading or unloading can be done in ten minutes, and cranes or other expensive freight handling equipment are not needed.

Another idea for reducing costs and getting freight traffic off the roads is the split-ship. This is designed to be narrow enough to travel down canals, yet wide enough to be seaworthy. Each half of a split ship can carry up to 550 tonnes, or the equivalent of about 200 lorry loads. The vessel is split for canal travel, but both sides meet up and join together to form a single stable ship for sea journeys. Freight could travel by this method through Britain's canal system to the open sea, then on through Europe's canal system to the upper reaches of the River Seine in France, the River Wista in Poland and the River Rhine in Germany. It has not been built yet, but a Danish company has expressed interest in developing the idea.

△ *The split-ship is still on the drawing-board.*

WHAT IS IT?

In transport design, there will always be room for the individual, the dreamer. It seems unlikely that many of these vehicles will actually become familiar, everyday methods of transport – but anything is possible!

The designer of the flying car below, a former professor of aeronautics at the University of California, thinks it will be a common sight in the skies by early next century.

The M400 is about the size of a large estate car. When the vehicle is travelling on the road it can go at about 110 kph (70 mph), but when it is airborne it should reach a top speed of 650 kph (400 mph). Eight small petrol-powered rotary engines in four ducts give it vertical take-off ability, so it needs no runway. When it is in the air, the direction of thrust from the engines is gradually changed from the vertical to the horizontal by adjusting a series of vanes in the duct exit. While it is cruising, the vehicle depends like a plane on the lift generated by its aerodynamic shape, and is intended to reach a height of 9,450 m (31,000 ft). The designer claims that maintenance costs will be low because the total number of moving parts, including the eight rotary engines, is less than that found in a conventional car. He also believes that, if the cars are driven by computers linked to satellite air navigation systems, then thousands of these craft could take to the air without risk of accident.

△ The prototype M400. The inventor hopes it will be airborne soon. Earlier designs have already flown.

If you like the idea of flying faster than sound in a plane you have built yourself, then an American company hopes to be able to sell you this DIY supersonic jet aircraft for around £150,000. The kit includes an engine capable of Mach 1.4. The designer claims it can be put together by any competent amateur and flown by anyone with an ordinary private pilot's licence. However, the company has yet to get final approval to market the plane!

△ The two-seater BD-10 supersonic jet comes as a kit. Fully assembled, it should weigh less than 5,443 kg (12,000 lb), so no special pilot's licence is needed.

ROAD FACTS

- The first motorway was built in 1924 from Milan to Varese in Italy.

- The shortest street in the world is Elgin Street in Bacup, Lancashire, England – a mere 5.18 m (16.99 ft) long.

- The longest traffic jam in the world was in France in 1980. It stretched 176 km (110 miles) between Paris and Lyons.

- The most number of cars involved in a jam was in 1990 with around 1.5 million cars at the East-West German border.

- The cats-eyes road-marking system was invented in Britain in 1934.

IN THE AIR

- The first controlled and sustained powered flight was on 17 December 1903 by Orville Wright in the *Flyer* designed with his brother Wilbur.

- The first trans-Atlantic flight was in May 1909 by Albert Cushing Read and his crew in a US Navy Curtiss flying-boat. The journey took 15 hours 18 minutes and averaged 151.4 kph (94.6 mph).

- The first flight across the English Channel was made by Louis Blériot on 25 July 1909. The journey took 36.5 minutes.

- The jet engine was patented in the United Kingdom in 1930.

- The record flying time for a commercial plane from New York to London was by Concorde in 2 hours 55 minutes 15 seconds in April 1990.

- The fastest journey around the world by a commercial plane was an Air France Concorde, which did the round trip in October 1992 in 32 hours 49 minutes, with six refuelling stops.

- The first non-stop unrefuelled flight around the world was by Dick Rutan and Jeana Yeager in 1986 in the plane *Voyager* designed by Dick's brother Bert Rutan. The journey took 9 days 3 minutes 44 seconds.

△ *The record-breaking plane,* Voyager.

- The biggest commercial plane is the Boeing 747-400, the Jumbo, which can squeeze in 570 passengers. It has a wingspan of 64.44 m (211.41 ft) and is 70.51 m (231.32 ft) long.

- The biggest transport plane is the Russian Antinov An225 124 which has a maximum take-off weight of 508 tonnes.

- The smallest plane ever flown is the *Bumble Bee Two*, with a wingspan of 1.68 m (5.51 ft), and only 2.64 m (8.66 ft) long. Designed and built by Robert Starr of Arizona, the *Bumble Bee Two* flew at 305.8 kph (191.1 mph) in May 1988.

△ *The aptly-named plane* Bumble Bee Two.

- The largest wingspan was 97.51 m (319.91 ft) on Howard Hughes' flying-boat *Spruce Goose*. It flew only once in 1947.

- The first automatic landing by a scheduled passenger airline was made by a Trident in 1967 at London's Heathrow airport, using an ILS (Instrument Landing System).

ON THE RAILS

- The first steam engine was demonstrated by Richard Trevithick in 1804 in South Wales.

- The fastest steam train was the British *Mallard* which went at 202.8 kph (126.7 mph) in July 1938.

- The last steam engine was built in China in 1988.

- The first electric locomotive was demonstrated in Berlin in 1879.

- The fastest passenger train in the world is the French TGV (Train à Grande Vitesse). It went at 515 kph (320 mph) between Courtlain and Tours on 18 May 1990.

- The fastest train using magnetic levitation was in 1979 when the Japanese train achieved 517 kph (323 mph).

- The world's longest railway track is the Trans-Siberian, which runs for 9,438 km (5,865 miles) from Moscow in Russia to Nakhoda on the east coast. The journey takes 8 days 4 hours 25 minutes.

- The railway network in China is expanding at 1,000 km (600 miles) a year.

IN THE CITY

- The greatest density of traffic is in Hong Kong: in 1991 there were 379,697 vehicles on 1,529 km (950 miles) of road. That is 4.02 m (13.2 ft) per car.

- The highest number of Rolls Royces *per capita* is in Hong Kong which also has the largest number of Mercedes Benz cars outside Germany.

- The first traffic lights were used in the USA in 1914.

- The first parking meters were installed in Oklahoma City, USA on 19 July 1935.

- The first wheel clamp was called the Denver Boot because it first appeared in the Colorado city of Denver in 1949.

- The longest escalator in the world is 800 m (2,625 ft) long and carries commuters between the Peak (a mountain on the island of Hong Kong) and the centre of the city.

HUMAN POWER

- The number of bicycles worldwide is estimated to be 800 million; half of these are in China where 1 person in 3 owns a bike.

- The bicycle has remained virtually unchanged since the British Rover safety cycle was built in 1885 by John Starley.

- The fastest speed for a one-hour cycle ride is held by Pat Kinch on the *Kingcycle Bean*. In 1990, he averaged 75.57 kph (46.96 mph).

△ Pat Kinch on the Kingcycle Bean.

- The first human-powered flight was in 1977 when British designer Paul MacCready's aircraft *Gossamer Condor* flew 1.6 km (1 mile) over a special course.

- The first human-powered flight across the English Channel was in 1979 by Bryan Allen in the *Gossamer Albatross* designed by Paul MacCready. He crossed in 2 hours 49 minutes.

- The longest human-powered flight was from Crete to the island of Santorini on 23 April 1988. The *Daedalus*, designed and built by the Massachusetts Institute of Technology, was flown by Knellos Kanellopoulos who completed the 119 km (74 miles) at an average speed of 29.7 kph (18.6 mph).

AT SEA

△ *Springwalker, a different way to travel.*

- A bicycle with legs is the idea of Californian inventors John Dick and Bruce Crapuchettes. Pumping the pedals of the Springwalker up and down transmits power through a system of pulleys and levers to two long legs. The inventors hope to be able to travel at 40 kph (25 mph).

- The hovercraft was invented by Christopher Cockerell in 1955 and first came into service in 1962 between Rhylin, North Wales and Wallasey, Cheshire in England. It started regular service across the English Channel in 1968.

- A journey around the world in 84 days underwater was achieved by the American nuclear submarine *Triton* in 1960.

- The world speed record was set by Kenneth Warby on 8 October 1978 with a speed of 277.57 knots (514.2 kph or 319.627 mph) in the hydroplane *Spirit of Australia* on Blowering Dam Lake, Australia.

- The fastest crossing of the English Channel by commercial ferry was on 9 February 1982 when the *Pride of Free Enterprise* crossed from Dover to Calais in 52 minutes 49 seconds - there was a Force 7 gale blowing at the time!

- The fastest crossing of the English Channel by hovercraft was on 1 September 1984 when the *Swift* crossed from Dover to Calais in 24 minutes 8.4 seconds.

- The record for the fastest Atlantic crossing by boat is held by Hoverspeed's SeaCat *Great Britain*, which did the journey at an average speed of 36.966 knots in June 1990.

- The largest merchant fleet in the world in 1990 was under the flag of Liberia.

△ *The hydroplane,* Spirit of Australia.

INDEX

aerodynamics 6, 16, 31, 36, 42
airbags 14, 15
alternative fuels 6–7
Apollo 9
automated bus stops 29

batteries 9, 31
beacon systems 10, 32
bicycles 30–31
   electric 31
   Lotus 31
   magnesium 30
black box 15
Boeing 747 16
Boardman, Chris 31
Bullet train 24
bus-ways 29

car convoys 11
Caspian sea monster 37
catamarans 36
Channel Tunnel 38–39
computer operated transport 4, 14, 23, 28, 29
Concorde 20, 21

DIY plane 42–43

electric cars 9, 12, 31
   Zoom 33
electronic guide systems 10

flying car 42, 43
freight 40–41
fuel efficiency 4, 6, 16, 34, 41

greenhouse effect 8
Ground Proximity Warning System 22

'hybrid' cars 12-13
hydrofoils 36
hydrogen fuel 9, 21
hypersonic transports 21

Il Pendelino 25
ILS 22

jet engines 21

laminar flow 16
lean-burn engines 6
light railway systems 28

magnetic drive 35
magnetic levitation 4, 26-27
magnetic ships 35
methane 8
methanol 8, 12
MLS 22
Moy, Thomas 36

pedestrianization 33
'People Movers' 29
petrol 6, 12
pollutants 8, 20
pollution 4, 6, 9, 16, 32, 40
propfans 17
public transport 28-29

railway network, European 24-25
ramjet engines 21
recycling 6

safety 14–15, 22–23, 25, 32
Sinclair, Clive 31
'smart' cars 10, 11
split-ships 41
sprinkler systems 18–19
Super-Concordes 20–21
super jumbos 18–19

tiphook trailers 41
traffic congestion 4, 10, 16, 28
Trains à Grande Vitesse 24, 38
Transrapid 06 26–27
turbofans 17
two-stroke engines 7

underground systems 28, 29
unmanned ships 35

variable bypass ducting 21

wing drag 16
wing-in-ground effect 37
Wingsails 34